ENVIRONMENTS
of Our
EARTH

ENVIRONMENTS
of Our
EARTH

By **Lawrence F. Lowery**

Illustrated by Linda Olliver

NSTA **Kids**
National Science Teachers Association
Arlington, Virginia

Claire Reinburg, Director
Jennifer Horak, Managing Editor
Andrew Cooke, Senior Editor
Amanda O'Brien, Associate Editor
Wendy Rubin, Associate Editor
Amy America, Book Acquisitions Coordinator

ART AND DESIGN
Will Thomas Jr., Director
Joseph Butera, Cover, Interior Design
Illustrations by Linda Olliver

PRINTING AND PRODUCTION
Catherine Lorrain, Director

NATIONAL SCIENCE TEACHERS ASSOCIATION
David L. Evans, Executive Director
David Beacom, Publisher

1840 Wilson Blvd., Arlington, VA 22201
www.nsta.org/store
For customer service inquiries, please call 800-277-5300.

NSTA is committed to publishing material that promotes the best in inquiry-based science education. However, conditions of actual use may vary, and the safety proce-dures and practices described in this book are intended to serve only as a guide. Additional precautionary measures may be required. NSTA and the authors do not war-rant or represent that the procedures and practices in this book meet any safety code or standard of federal, state, or local regulations. NSTA and the authors disclaim any liability for personal injury or damage to property arising out of or relating to the use of this book, including any of the recommendations, instructions, or materials contained therein.

Library of Congress Cataloging-in-Publication Data
Lowery, Lawrence F.
  Environments of our earth / by Lawrence F. Lowery ; illustrated by Linda Olliver.
     pages cm. -- (I wonder why)
  Audience: K to grade 3.
  ISBN 978-1-938946-15-8 -- ISBN 978-1-938946-69-1 (e-book) 1. Biotic communities--Juvenile literature. 2. Ecology--Juvenile literature. I. Olliver, Linda, illustrator. II. Title.
  QH541.14.L69 2013
  577.8'2--dc23
                        2013021580

Cataloging-in-Publication Data are also available from the Library of Congress for the e-book.

Introduction

The *I Wonder Why* books are science books created specifically for young learners who are in their first years of school. The content for each book was chosen to be appropriate for youngsters who are beginning to construct knowledge of the world around them. These youngsters ask questions. They want to know about things. They are more curious than they will be when they are a decade older. Research shows that science is students' favorite subject when they enter school for the first time.

Science is both *what* we know and *how* we come to know it. What we know is the content knowledge that accumulates over time as scientists continue to explore the universe in which we live. How we come to know science is the set of thinking and reasoning processes we use to get answers to the questions and inquiries in which we are engaged.

Scientists learn by observing, comparing, and organizing the objects and ideas they are investigating. Children learn the same way. These thinking processes are among several inquiry behaviors that enable us to find out about our world and how it works. Observing, comparing, and organizing are fundamental to the more advanced thinking processes of relating, experimenting, and inferring.

The five books in this set of the *I Wonder Why* series focus on Earth science content. The materials of our Earth are mostly in the forms of solids (rocks and minerals), liquids (water), and gases (air). Inquiries about these materials are initiated by curiosity. When we don't know something about an area of interest, we try to understand it by asking questions and doing investigations. These five Earth science books are written from the learner's point of view: *How Does the Wind Blow?*; *Clouds, Rain, Clouds Again*; *Spenser and the Rocks*; *Environments of Our Earth*; and *Up, Up in a Balloon*. Children inquire about pebbles and rocks, rain and wind, and jungles and deserts. Their curiosity leads them to ask questions about land forms, weather, and climate.

The information in these books leads the characters and the reader to discover how wind can be measured and how powerful it can be, how the water cycle works, that living things need water to survive, and that plants and animals have adapted to different climate-related environments. They also learn how people have learned to fly in the ocean of air that surrounds Earth.

Each book uses a different approach to take the reader through simple scientific information. One book is expository, providing factual information. Several are narratives that allow a story to unfold. Another provides a historical perspective that tells how we gradually learn science through experimentations over time. The combination of different artwork, literary perspectives, and scientific knowledge brings the content to the reader through several instructional avenues.

In addition, the content in these books correlates to criteria set forth by national standards. Often the content is woven into each book so that its presence is subtle but powerful. The science activities in the Parent/Teacher Handbook section in each book enable learners to carry out their own investigations that relate to the content of the book. The materials needed for these activities are easily obtained, and the activities have been tested with youngsters to be sure they are age appropriate.

After students have completed a science activity, rereading or referring back to the book and talking about connections with the activity can be a deepening experience that stabilizes the learning as a long-term memory.

The surface of our Earth is made up of water and land.
There is more water on our Earth than there is land.
Rain brings much of the water to the land, but some parts
of our Earth get more rain than other parts.

Living things need water to live.

Many living things grow well where there is a good supply of water. Fewer living things grow where there is very little water.

The places where living things live are called environments.

One of the wettest environments on our Earth is
Mount Waialeale on the island of Kauai in Hawaii.

It rains on Mount Waialeale nearly every day.
Mount Waialeale gets nearly 500 inches
of rainfall every year.

Environments on our Earth that have a great amount of rain all year long are called tropical rain forests.

Many types of plants—such as shrubs, herbs, and vines—grow close together in tropical rain forests.

Some plants grow very large and tall. Tall, broad-leaf evergreen trees branch out and cover the rain forest like a canopy.

Many types of animals—such as colorful birds, reptiles, monkeys, and insects—live in the thick rain forests.

There is no winter in a tropical rain forest.

These warm, wet environments can be found near Earth's equator in places such as South America, Africa, and Indonesia.

Tropical
Rain Forests

A rainy but cooler place is called a woodland environment.

In a woodland environment, large trees grow close together.

Ferns, mosses, and mushrooms grow on the woodland floor among the tall trees.

Wolves, foxes, bears, and squirrels build their homes in thick woodland environments.

Some of our Earth's woodland environments are found in North America, Asia, Africa, Spain, and Chile.

Woodlands

N
W E
S

Savannas are environments that have much less rain than woodland environments. Savannas have a short dry period each year.

On savannas, there is much more grass growing than other plants and trees.

On savannas, trees do not grow very tall.
They usually grow alone or in small groups
that seem to dot the grasslands.

Some animals that live in the savanna environment are leopards, cheetahs, elephants, giraffes, lions, and zebras.

Unless paths have been made by people or animals, it is difficult to walk through the thick, coarse grasses of the savanna.

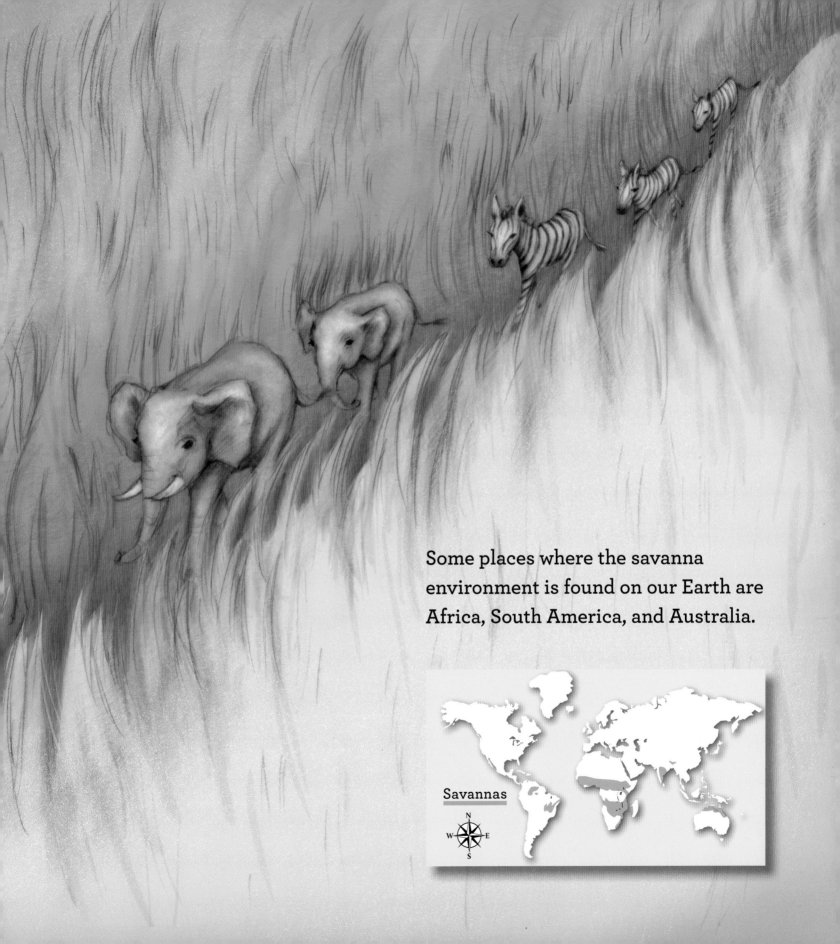

Some places where the savanna environment is found on our Earth are Africa, South America, and Australia.

Savannas

N
W E
S

Prairies are another kind of grassland.

The tall grasses of the prairies get much less water than the tall grasses in the savannas. Most of the rainfall on the prairies comes during the summer season.

Even though there is enough water for grasses, herbs, and shrubs to grow, trees seldom grow in prairie environments. Nobody knows why.

Some animals that live in the prairie environment are the badger, bobcat, bumblebee, fox, coyote, prairie dog, and American bald eagle.

In the spring and summer, flowering plants bloom among the grasslands in the prairies.

At this time of the year, the American prairie looks like a colorful flower garden.

Some prairie environments can be found in Argentina, Brazil, Canada, Mexico, and the United States.

Prairies

An environment that gets less rain than a prairie
is another kind of grassland called a steppe.
A small amount of rain falls on the steppe in the
late spring and early summer.

On a steppe, the grasses are shorter and make the
land look like it is covered with an endless carpet.

Cattle, sheep, and horses graze on the steppe.

At one time, great herds of bison wandered over this kind of grassland in the United States.

Some steppe environments can be found in China, South America, and the United States.

Steppes

The desert environment is a place where few plants can grow because of the small amount of rainfall.

Where there is so little rain, many desert plants—such as sage and cactus—store water in their thin stems.

The horned lizard, iguana, kit fox, and pack rat are some of the animals that live in the desert environment. Most do not drink water. They get water from eating plants and seeds that store water.

Many desert animals live underground during the day to stay out of the heat.

Some deserts are almost completely without water. On such deserts, winds sometimes blow the sand to create small hills or dunes.

One of the driest places on Earth is the Atacama Desert in Chile. It rains so little there that the rainfall usually cannot be measured.

On our Earth, very large deserts can be found in North America, South America, Africa, Australia, and the Antarctic.

Deserts

Water is an important part of every environment in which plants and animals live. Some environments have great amounts of water, while some have very little.

Plants and animals have adapted to be able to live in these different environments.

People can help preserve the different environments so the plants and animals that live in them will continue to be on our Earth for a long, long time.

ENVIRONMENTS of Our EARTH

PARENT/TEACHER HANDBOOK

Introduction

Environments of Our Earth is about how regions on Earth are shaped by rainfall. The Earth's atmosphere has regions characterized by large-scale rising air and other regions with large-scale descending air; these regions vary by latitude and season. The global distribution of rainfall shows that the wettest areas are in regions with rising air, while the driest areas (deserts and polar areas) are in regions with descending air. Due to this difference, rainfall is a major factor that distinguishes environments on Earth from one another, ranging from the very wet and humid conditions of the tropics to desert regions where the amount of rainfall is almost too small to measure.

One example of an area that receives frequent rain is Mount Waialeale on Kauai, Hawaii. Mount Waialeale is a high peak directly in the path of steady trade winds that carry a lot of moisture all year round. On the upwind (northeast) side of Mount Waialeale, the air rises and condenses, resulting in almost constant clouds. It rains there nearly every day. Mount Waialeale is among the wettest places on Earth: Over a 10-year period, annual rainfall averages close to 5 feet. In contrast, on the downwind (southwest) side of Mount Waialeale, the air descends, warms, and is dryer. The result is a semi-arid area with a little more than 1 foot of rain per year. This tremendous difference in precipitation occurs in a span of only 20 miles, and the plants and animals that are found in these two areas are quite different.

Inquiry Processes

Scientists sometimes seriate objects and ideas to better understand them. Rocks might be put in a seriated order from smallest to largest or heaviest to lightest. Regions of the world can be seriated on the basis of rainfall, from very wet regions with lots of rainfall to regions that seldom get a drop of rain. The amount of rainfall in a region is a major factor in that region's climate, and a seriated order of regions enables scientists to make comparisons of how regions are alike and different and better understand the factors that create the different regions.

Content

Climate classification systems such as the Köppen climate classification system use average annual rainfall to differentiate between differing climate regions. Rain gauges are used to gather and measure rainfall.

The major cause of rain production is moisture-filled air. If enough moisture and upward motion of the air are present, precipitation falls from clouds. Land areas near large bodies of water (oceans and seas) receive more rain than inland regions because the winds lose moisture as they move upward over the land and become quite dry by the time they reach the interior of a continent.

Rainfall measures are one of the primary elements that determine climatic conditions, and they are a factor of tremendous importance in the distribution of plant and animal life on Earth. Organisms that live in the different environmental regions, from wet to dry, adapt to survive in those regions. For example, leaves on plants in the humid, wet tropics tend to be large and broad. The leaves release water that the plant does not use back into the air. The "needles" on cacti are actually leaves that have adapted for survival in dry environments. The needle-like leaves do not release much water back into the air. Why do you think that is important for the plant's survival is a desert region?

Science Activities

Making a Rain Gauge

It is possible to measure the amount of rain that falls where you live. The rain gauge is an instrument used to measure rainfall. Any open container with straight sides can be used to measure amounts of rainfall. When the rain is collected, simply stick a ruler into the container to see how deep the water is. If the water is 1 in. (3 cm) deep, then 1 in. (3 cm) of rain has fallen.

Because most rainfall measures less than an inch, a true rain gauge is designed to catch a relatively wide area of rainfall and funnel it into a narrow area so that it will be deeper and can be measured more easily. (*Note:* Measurements must be taken soon after a rainfall, or evaporation will give inaccurate readings.)

To measure small amounts of rainfall, attach a test tube beside a strip of paper on a block of wood. Fill a wide-mouth, straight-sided jar with 1 in. (3 cm) of water. Pour this water into the test tube and mark the height on the strip of paper. Repeat this procedure using 3/4 in., 1/2 in., and 1/4 in., or use metric units for 2.5 cm, 2 cm, 1.5 cm, 1 cm, and 0.5 cm of water. Now the jar can be placed outdoors in the open. When rain is collected in the jar, pour it into the test tube to measure how much rain fell.

Building a Terrarium

The terrarium is a simulated environment designed for observing land plants and animals. Terrariums provide first-hand observations of tropical, temperate, and dry land areas and can include the plants and animals that have adapted to these land conditions. Here are two types of land conditions, each with a particular set of environmental characteristics:

The Woodland Environment

To create a temperate woodland environment, cover the bottom of a container with a layer of gravel for good drainage, then add several inches of fertile soil. Bury pieces of charcoal in the soil to absorb gases and prevent the soil from souring.

Add a few rocks and a dish of water to provide the necessary moisture for this environment. With a glass top on the container, the terrarium will maintain it own water cycle. Large glass jars or an aquarium tank make fine containers.

Mosses, ferns, and other small woodland plants can be placed in the container and watered infrequently. Too much water will cause mold to appear. If this happens, remove the moldy plants and leave the top off the container to allow some of the excess moisture to evaporate. Small animals such as frogs, turtles, toads, salamanders, newts, and snakes can be added. (*Note:* Animals must be compatible with one another, and you should not put in too many. Information about appropriate food for the animals you select can be found on the internet. Be sure to remove any excess food the animals do not eat.)

The Desert Environment

To create a desert environment, build a base in a container with 2 in. (5 cm) of gravel and coarse sand. Make the base slightly lower at one end of the container. Add rocks and a small dish of water. Plant desert plants such as cacti. Small desert snakes, lizards, and horned toads can be kept in this environment. (*Note:* Desert plants decay from too much water, and molds might grow. Be sure to remove them if this happens.) This terrarium does not need a cover, although mesh screening can be used. You can put between two and four fluorescent or 60 W bulbs near the terrarium to provide variations in warmth.

After you have created a terrarium, you can test the influence of changes in light or temperature on the environment. For example, you could try putting the terrarium in different places, such as in darkness or in direct sunlight, or you could turn the container partway around for a few days and observe what happens to the plants and animals. You could also try setting a heat source at one end of the container to see what happens. Keep a record of your findings.